Longman Practice Exam Papers

A-level Biology

Chris Millican

Series editors:

Geoff Black and Stuart Wall

Titles available for A-Level

Biology

Business Studies

Chemistry

Physics

Psychology

Pure Mathematics and Mechanics

Pure Mathematics and Statistics

Addison Wesley Longman Limited
Edinburgh Gate, Harlow
Essex CM20 2JE, England
and Associated Companies throughout the World

© Addison Wesley Longman 1999

The right of Chris Millican to be identified as author of this work has been
asserted by her in accordance with the Copyright, Designs and Patents Act 1988.

First published 1999

ISBN 0-582-36919-3

British Library Cataloguing in Publication Data
A catalogue record for this book is available from the British Library

Set in Times 11/13 and Gill Sans by 38

Printed in Singapore

Contents

Introduction

Organisation of your course and assessment

Many students follow a modular course at A-level, where aspects of the syllabus are assessed during the course. Unfortunately a particular topic, for example digestion, will be placed in different modules by different exam boards. The module exams may vary both in length, and in the number of marks available. It is therefore vital that you have the following information:

- the topics that are included in a particular module (ask your teacher for a syllabus or topic list if you do not already have one)
- the duration of your module tests
- the number of marks available for each module test
- the types of questions you will be expected to answer – for example, short answer, graph drawing, essays, and so on.

How to use this book

There are four separate practice exam papers in this book.

Papers 1–3 contain short answer, data analysis and longer prose questions. Each paper lasts 2 hours and is worth 100 marks. All questions are compulsory.

Paper 4 is an essay paper. There is a choice of questions, and you are required to answer three. The paper lasts for 2 hours and is worth 60 marks.

In order to get the most out of these practice exam papers you should:

- make sure that you have a quiet place to work where you will be undisturbed for the duration of the practice exam
- make sure that you have 2 hours to complete the practice exam; keep a watch or clock in view
- have all the equipment that you need, such as pens, pencils, ruler, calculator (with new or spare battery), etc.
- answer all questions in the spaces provided except longer prose answers and essays, which should be written on ruled paper
- use the solutions at the back of this book (pages 40–58) to mark your work, or get a friend to mark it for you. Remember that you are unlikely to use exactly the same words or phrases as you will find in the mark scheme, but examiners often look for key words in an answer; including these will help you to gain higher marks

- do not be too generous when awarding yourself marks. A-level examiners will stick very closely to the mark scheme they are given
- correct any mistakes and work out why you lost marks; this is an important part of improving your performance
- use your mark to work out an approximate A-level grade (page vi will help you to do this).

How to achieve high marks

There are two criteria for achieving high marks:

- know your work well
- communicate this knowledge clearly in the time available

The second criterion is sometimes called 'exam technique', and this book will help you to improve your technique, if you consider the following advice.

Short-answer questions

- Look at the space provided for your answer, and the number of marks available; this will indicate how detailed your answer should be.
- Look at how the question is phrased, and make sure that you do what is required; for example:

Describe	Give a brief account of events, or translate data (in a table or graph) into words. An explanation is not needed.
Explain	Use biological theory knowledge to give reasons for what is happening.
Comment on	Usually a combination of *describe* and *explain*; that is, say what is happening and why.
Suggest	Put forward ideas to explain; there are usually several correct alternative answers to a question phrased like this.

- Try to include biological terms wherever possible.
- When drawing graphs always label the axes and choose an appropriate scale. Plot points neatly and accurately, and join them with a smooth curve or line of best fit.
- When carrying out calculations, always include units in your answer and show all your working.

Longer prose answers

- These are worth up to 10 marks. You should aim to write about one and a half sides, but do not repeat yourself needlessly.
- They are marked according to a rigid mark scheme, in which one statement gains 1 mark.
- There is no need to write in formal essay style, that is, with an introduction, separate paragraphs for development of ideas, and so on.
- Diagrams may be included if they are fully labelled and clarify your answer, but they should not form the main part of your answer.
- You should spend approximately 20–25 minutes on these questions.

Essays

- These are usually worth 20–25 marks. You should aim to write about three sides, but do not repeat yourself needlessly.
- They are not marked according to a rigid mark scheme, because there are so many ways of writing a good essay. However, you will normally be given marks in three categories:

Scientific knowledge
This is how the majority of the marks are awarded. Consider these two statements:

'ATP is made in respiration.'
'ATP is made during the third stage of respiration, the electron transport system. As electrons are passed from one carrier to another, energy is released and this is used to convert ADP into ATP in a process known as oxidative phosphorylation.'

Both of these statements are true, but the second would gain far more credit as it includes much more scientific information.

Breadth

This is how well you cover the different aspects of the essay set. If the title was 'Give an account of gas exchange', you would gain a low mark for breadth if you wrote only about humans. You should include information about other animals, such as fish, insects and plants, to gain high marks.

Style

You should structure your essay carefully so that there is a brief introduction followed by a number of development paragraphs, ending with a suitable closing paragraph.

The points should flow in a logical order and be expressed in clear, grammatically correct terms. Spelling and punctuation should be of a high standard.

■ Plan your essay carefully. You should spend at least 10 minutes on this, because good planning is vital if you are to achieve high marks.

■ Most exam boards do not allow diagrams in essay answers. Check with your teacher whether you are allowed to include them.

Editors' preface

Longman Practice Exam Papers are written by experienced A-level examiners and teachers. They will provide you with an ideal opportunity to practise under exam-type conditions before your actual school or college mocks or before the A-level examination itself. As well as becoming familiar with the vital skill of pacing yourself through a whole exam paper, you can check your answers against examiner solutions and mark schemes to assess the level you have reached.

Longman Practice Exam Papers can be used alongside *Longman A-level Study Guides* and *Longman Exam Practice Kits* to provide a comprehensive range of home study support as you prepare to take your A-level in each subject covered.

Acknowledgements

I would like to thank the following people for their help: Jonathan Bow, Lloyd Evans, Emma Hedges and Salonee Patel for permission to use experimental data; Anne-Marie Crocombe, David Gore and Richard Hughes for trialling the questions in this book.

I would also like to thank Clive Hurford for his help with the preparation of the manuscript, and his support as always.

Chris Millican

How well did you do?

How to analyse your mark

Papers 1–3 are each worth 100 marks. Paper 4 is worth 60 marks.
Add up your marks for each paper, and calculate the percentage marks.

It is not possible to give the exact mark needed for a particular grade, as these are worked out by the exam boards and will vary slightly from year to year. The following grade boundaries give an indication of the standard needed to achieve each grade:

A   80%+
B   70–79%
C   60–69%
D   50–59%
E   40–49%
N   30–39%
U   up to 29%

Learning from your mistakes

There are four main reasons why you may have achieved a low score in the exam:

- *You did not know your work.* The solution to this problem is easy – you need to revise more carefully, then try more practice papers.
- *You misread the question, so you gave the wrong information in the answer.* In future, underline key words in the question, and read the whole question before you start to answer.
- *You knew the answer, but did not include enough information to achieve full marks.* Look at the number of lines provided and the number of marks available for each part of the question; Underline the key words for each answer in the mark scheme.
- *You ran out of time, and did not finish the paper.* Each exam lasts 2 hours, so you should check your progress after 30 minutes, and 1 hour. Essays and longer prose answers will always take longer than you think, but for other questions you should allow about 1 minute for each mark in the question.

If you did not achieve the grade you hoped for, use the points above to work out why – it will probably be a mixture of the reasons listed here. Once you know why you are losing marks you can start to improve your performance. Remember, everyone can improve their grade by working on their weak areas, and by practising plenty of questions from past papers.

Good luck!

Longman Examination Board

General Certificate of Education

A-level Biology

Paper 1

Time: 2 hours

Number	Mark
1.	
2.	
3.	
4.	
5.	
6.	
7.	
8.	
9.	
10.	
Total	

Instructions

- Attempt ALL the questions.

- Answer Questions 1–9 in the spaces provided on this exam paper. Answer Question 10 on separate ruled paper.

- Show all stages in any calculation, and state the units.

- Where diagrams are required, draw and label them clearly.

Information for candidates

- The marks available are shown in brackets after each question or part-question.

- This exam paper has 10 questions.

- You are allowed 2 hours for this paper.

- The maximum mark for this paper is 100.

---

1. The table below refers to different types of micro-organisms.

   If the statement is correct for a particular organism, place a tick in the appropriate box. If it is incorrect, place a cross in the box.

	Bacterium	Yeast	Virus
Has a cell wall			
Contains protein			
Contains DNA and RNA			
Contains mitochondria			
May have flagella			
Contains chloroplasts			

Leave margin blank

(Total 6 marks)

Turn over

1

2. The flow diagram shows some of the stages in genetic engineering to produce Human Growth Hormone (HGH). The diagrams are not drawn to scale.

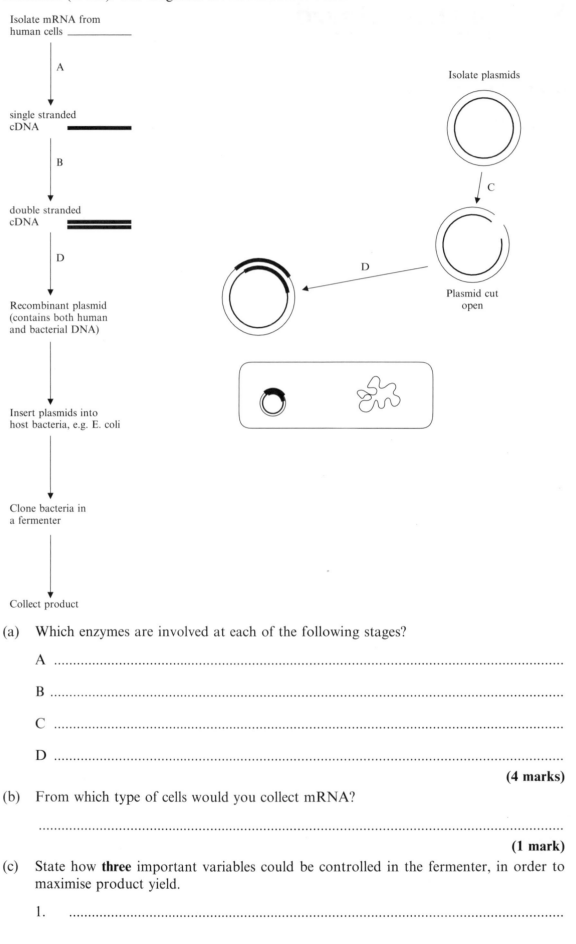

(a) Which enzymes are involved at each of the following stages?

A ....

B ....

C ....

D ....

(4 marks)

(b) From which type of cells would you collect mRNA?

....

(1 mark)

(c) State how **three** important variables could be controlled in the fermenter, in order to maximise product yield.

1. ....

....

Leave margin blank

2. ....

....

3. ....

....

(3 marks)

(Total 8 marks)

3. Read through the following account of photosynthesis, and complete the passage by writing the most appropriate word or words in the spaces.

Photosynthesis can be divided into two stages: the light dependent stage which occurs

in the .... of the chloroplast, and the light independent stage, which

occurs in the ....

In the first stage light energy is trapped by .... and electrons

are released. These are passed on to electron acceptors at a .... energy level

and as they are transferred through electron carriers energy is released. This is used to convert

.... to .... Electrons are lost from a .... molecule in

a process known as photolysis, and .... is a waste product of this reaction.

The hydrogen ions formed reduce the hydrogen carrier .... to ....

which is vital in the light independent stage. Here carbon dioxide combines with a 5-carbon

compound called .... in a reaction catalysed by the enzyme

.... **(Total 12 marks)**

4. The table below refers to some chemicals found in mammals. Complete the table by writing the names of the chemicals in the appropriate spaces.

Chemical	Function
	Causes ovulation to occur
	Raises blood sugar level by causing conversion of glycogen to glucose
	Acts as a neurotransmitter at the synapse
	A fall in blood levels of this chemical causes menstruation to start
	Causes uterine contractions and induces lactation
	The main neurotransmitter in the sympathetic nervous system

(Total 6 marks)

5. *Paramecium* are heterotrophic protoctists found in freshwater and soil. In 1934 Gause carried out an experiment to investigate competition between two species of *Paramecium*, called *Paramecium caudatum* and *Paramecium aurelia*, which is much smaller.

Turn over

In his first experiment, he grew the two populations separately and monitored population size over 20 days. In the second experiment, he grew the two populations together in the same culture vessel, and monitored the population sizes over 20 days. The results are shown in the graphs below.

Experiment 1
The two species are grown separately

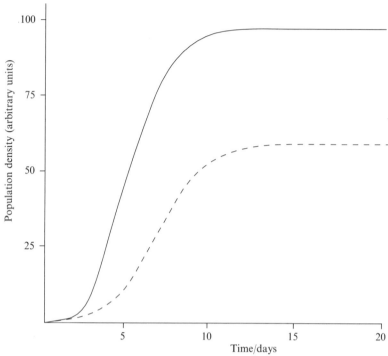

Experiment 2
The two species are grown together

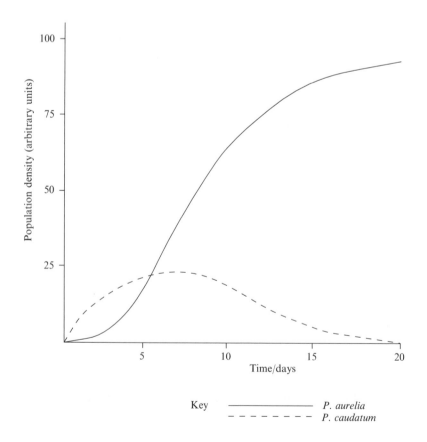

(a) (i) Comment on the growth curve for *P. aurelia* grown alone.

....

....

....

(2 marks)

(ii) Does this curve show density-dependent or density-independent growth? Give a reason for your choice.

....

....

....

(2 marks)

(b) Describe the population curves for the two species when they are grown together.

P. caudatum ....

....

P. aurelia ....

....

(4 marks)

(c) Explain the changes in population size for *P. caudatum*

(i) over the first 3 days ....

....

....

(1 mark)

(ii) towards the end of the experiment. ....

....

....

(1 mark)

(d) Comment on the final numbers of *P. aurelia* in both experiments.

....

....

....

....

....

(2 marks)

Turn over

(e) From experiments like this, Gause devised the competitive exclusion principle. State briefly what this says.

....

....

....

(1 mark)

(Total 13 marks)

6. A student investigated how the size of a population of yeast cells changed over 4 hours.

   She placed 5 g of yeast and 5 g of sugar in 50 ml of water in a conical flask. The flask was placed in a waterbath at 35 °C, and a sample of the mixture was removed every 20 minutes for counting.

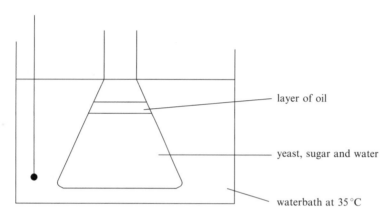

layer of oil

yeast, sugar and water

waterbath at 35 °C

   The sample was diluted by adding 1 ml of the yeast mixture to 99 ml of distilled water, then one drop was placed on a haemocytometer slide. When it was viewed under the microscope, it looked as shown in the figure below.

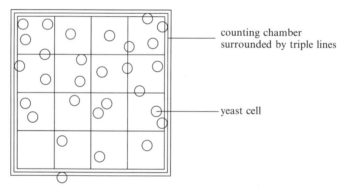

counting chamber surrounded by triple lines

yeast cell

   (a) With the coverslip in place, the counting chamber of the slide contains 2.5×10^{-4} ml of fluid.

   Counting only the cells totally within the chamber, and those touching or overlapping the top or right hand edges, calculate the number of yeast cells per ml of diluted sample. Show your working.

(3 marks)

(b)   Calculate the number of yeast cells per ml of undiluted sample. Show your working.

(1 mark)

The graph below shows how the number of yeast cells changed over the course of the experiment.

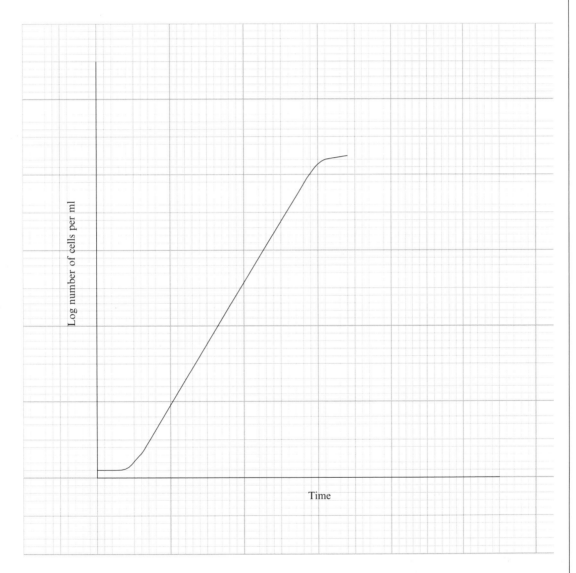

(c)   Why is the log number of cells plotted, rather than the actual number?

....

....

(1 mark)

Turn over

(d)   Give reasons to account for the shape of the graph.

....

....

....

....

....

....

....

(3 marks)

(e)   (i)   On the graph paper, sketch the shape of the line you would expect if the experiment had continued for a further 4 hours.   **(1 mark)**

(ii)   Give reasons to account for the shape of the line you have drawn.

....

....

....

(2 marks)

(Total 11 marks)

7.   The diagram below represents part of the process of protein synthesis (structures are not drawn to scale).

(a)   Which letter(s) represent each of the following? Each letter may be used once, more than once or not at all.

(i)   A structure made mostly of phospholipid. ....

(ii)   A structure containing ribose. ....

(iii) The site of a recent condensation reaction. ....

(iv) A structure containing an anticodon. ....

(v) A structure containing uracil. .... **(5 marks)**

(b) Describe the translation stage of protein synthesis.

....

....

....

....

....

(4 marks)

(Total 9 marks)

8. The graph below shows changes in pressure within the heart during the cardiac cycle.

(a) Use the graph to calculate the heart rate in beats per minute.

....

(1 mark)

(b) Explain why pressure in the ventricles begins to rise at point A.

....

....

....

(2 marks)

Turn over

(c) (i)   At what time do the aortic valves close? ....

(1 mark)

(ii)   Give a reason to explain your answer to (c)(i).

....

....

(1 mark)

(d) Why does blood pressure in the aorta fall slowly once the ventricle has stopped contracting?

....

....

(1 mark)

(e) Heart rate is controlled by the cardiac centre of the brain.
In which part of the brain is the cardiac centre found?

....

(1 mark)

(f) (i)   Mark the position and label each of the following on the diagram of the heart below.

Sino-atrial node
Atrio-ventricular node
Purkinje fibres

(3 marks)

(ii)   Explain the importance of each of those structures in the cardiac cycle.

....

....

....

....

....

(3 marks)

(Total 13 marks)

9. A student carried out an experiment to determine the effect of sodium chloride on the enzyme amylase.

1 ml of amylase was mixed with 1 ml of sodium chloride solution, then 5 ml of 1% starch solution was added. Each minute a drop of this mixture was tested with iodine solution, and the colour change was recorded. This continued until there was no change in colour – this was recorded as the achromatic time.

This procedure was carried out at a range of concentrations of sodium chloride, with each concentration being tested twice. The table below shows the results obtained.

Concentration of sodium chloride solution (%)	Achromatic time (s)		Mean achromatic time (s)	Rate of reaction (s^{-1})
	1	2		
0	20	20	20	0.05
2	19	19		
4	18	19		
6	17	18		
8	15	15		
10	14	13		
12	12	12		
14	12	11		
16	10	10		
18	9	10		

(a) Complete the table by calculating the mean achromatic times and rates of reaction. The first pair has been done for you. **(2 marks)**

(b) Using the grid below, plot a graph to show how rate of reaction varies with the concentration of sodium chloride solution added.

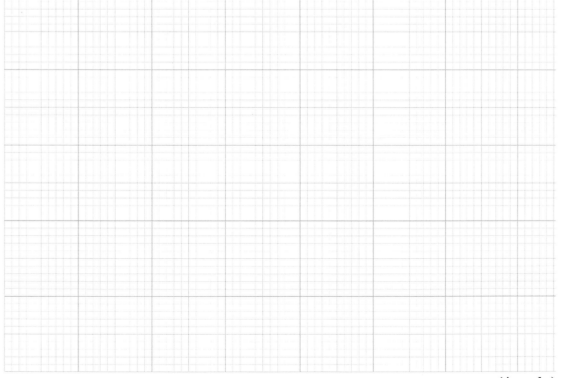

(4 marks)

Turn over

(c) Write a suitable conclusion for this experiment.

....

....

(1 mark)

(d) From your knowledge of enzyme activity, suggest why sodium chloride may have had this effect.

....

....

....

(2 marks)

(e) When the student tried the experiment with a 30% sodium chloride solution, the achromatic time was 37 minutes. Suggest why this may have occurred.

....

....

(1 mark)

(f) Suggest **two** ways that this experiment could be improved.

....

....

(2 marks)

(Total 12 marks)

10. **Answer the following question on separate ruled paper.**

Give an account of the ultrastructure and mechanism of contraction of striated muscle.

(Total 10 marks)

Total marks for paper: 100

Longman Examination Board

General Certificate of Education

A-level Biology

Paper 2

Time: 2 hours

Number	Mark
1.	
2.	
3.	
4.	
5.	
6.	
7.	
8.	
9.	
10.	
Total	

Instructions

■ Attempt ALL the questions.

■ Answer Questions 1–9 in the spaces provided on this exam paper.
Answer Question 10 on separate ruled paper.

■ Show all stages in any calculation, and state the units.

■ Where diagrams are required, draw and label them clearly.

Information for candidates

■ The marks available are shown in brackets after each question or part-question.

■ This exam paper has 10 questions.

■ You are allowed 2 hours for this paper.

■ The maximum mark for this paper is 100.

---

1. The table below refers to some biologically important molecules. Complete the table by adding ticks or crosses to show whether the molecules contain the elements listed.

	Nitrogen	Carbon	Iron
Glucose			
Glycine (amino acid)			
Chlorophyll			
Triglyceride			
ATP			
Haemoglobin			

Leave margin blank

(Total 6 marks)

Turn over

13

2. (a) The diagram below shows a chloroplast. Use the letters given in the diagram to answer the following questions; each letter may be used once, more than once or not at all.

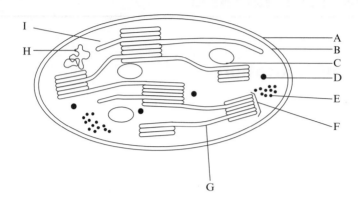

In which part of the chloroplast

(i)   is starch stored? .... **(1 mark)**

(ii) is photosystem I found? .... **(1 mark)**

(iii) does the light independent stage of photosynthesis occur? .... **(1 mark)**

(iv) is ATP made? .... **(1 mark)**

(b) The graph below shows how the rate of photosynthesis varies under different conditions.

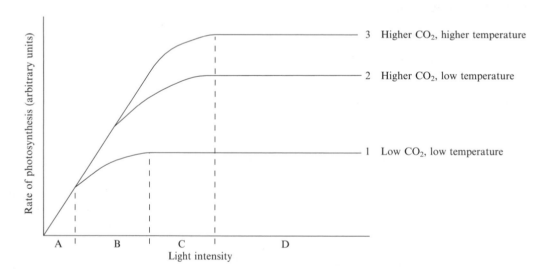

Using only evidence from the graph, mark each of these statements **true** or **false**.

(i)   There is a directly proportional relationship between light intensitity and

photosynthesis. .... **(1 mark)**

(ii) In curve 1, region B, carbon dioxide is the only limiting factor. .... **(1 mark)**

(iii) There is a directly proportional relationship between levels of carbon dioxide and

photosynthesis. .... **(1 mark)**

(iv) Temperature can act as a limiting factor in photosynthesis. .... **(1 mark)**

(c) The graph below shows the effect of light intensity on the amount of carbon dioxide taken up or released by a plant. Mark the following points clearly on the graph:

(i) the compensation point **(1 mark)**

(ii) light saturation **(1 mark)**

(iii) the region corresponding to no photosynthesis. **(1 mark)**

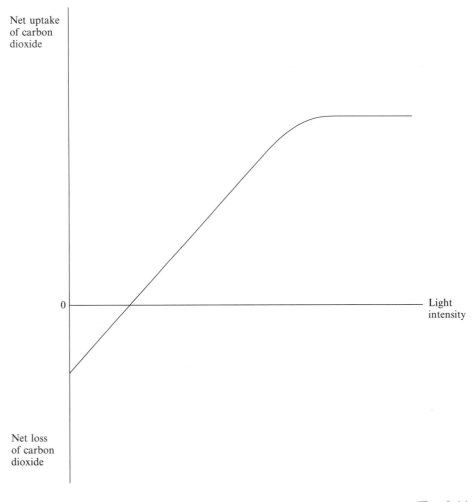

(Total 11 marks)

3. (a) The diagram below shows one of the stages of cell division during gamete production in a mammal.

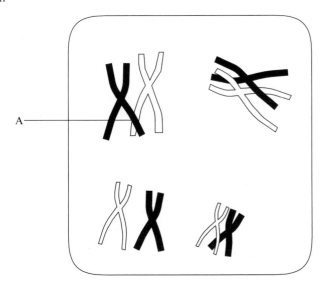

Turn over

(i) Which type of cell division is this? ....

(1 mark)

(ii) Which stage is shown? ....

(1 mark)

(iii) Give **two** reasons for your answer to (a)(ii).

....

....

(2 marks)

(iv) What is occurring at the point labelled A?

....

(1 mark)

(v) Explain how this will increase variation.

....

....

....

(2 marks)

(b) One method of artificial propagation in mammals uses the technique shown in the diagram below.

(i) Would the sheep formed as a result of this procedure be:

A genetically identical to sheep X
B phenotypically identical to sheep X
C genetically and phenotypically identical to sheep X
D different to sheep X?

(Answer A, B, C or D) ....

(1 mark)

Leave margin blank

(ii)   Give the reasons for your answer to (b)(i).

....

....

....

(2 marks)

(Total 10 marks)

4.   Peppered moths (*Biston betularia*) exist in two forms: light (speckled grey) and melanic (black).

In the 1950s Kettlewell bred large numbers of both the light and the melanic forms, and released equal numbers of the two forms in Birmingham and in Dorset. The table below shows the percentage of each type recovered later.

	% recovered at each location	
	Light	**Melanic**
Birmingham	13.1	27.5
Dorset	12.5	6.3

(a)   Which of the following statements best explains these results? Underline **one** answer.

The light moths have died more quickly in Dorset.

The light moths have bred more quickly in Dorset.

In Birmingham, light moths have developed melanic colouring.

In Birmingham, melanic moths survive longer than light moths.

In Dorset, melanic moths have become paler to blend in with their surroundings.

(1 mark)

(b)   Kettlewell interpreted these results in terms of a selective advantage. Explain how this could have occurred.

....

....

....

....

....

(4 marks)

(c)   This is an example of *polymorphism*. Explain the meaning of this term, giving another species as a named example.

....

....

....

(2 marks)

(Total 7 marks)

Turn over

5. A student wanted to investigate whether the amount of light available to a plant had an effect on leaf size. She measured the area of 20 leaves of *Stachys sylvatica* (Hedge woundwort) growing in full sun, and 20 leaves from *Stachys sylvatica* growing in shady conditions.

(a) Suggest how leaf area could be measured.

....

....

(1 mark)

(b) She plotted a histogram of her results, as shown opposite on page 19.

    (i) Suggest an explanation for the difference in leaf area of the two groups.

....

....

(2 marks)

    (ii) Apart from leaf area, state **one** other difference you would expect to find between the two groups of plants.

....

(1 mark)

(c) Name a statistical test you could carry out to determine if the two groups of plants had significantly different leaf areas.

....

(1 mark)

(d) The mean leaf area for plants in sun was $27.55\,cm^2$ with a standard deviation of $9.62\,cm^2$.

The mean leaf area for plants in shade was $55.5\,cm^2$ with a standard deviation of $13.4\,cm^2$.

What size range would include 95% of the plants in sun? Show your working.

(2 marks)

(Total 7 marks)

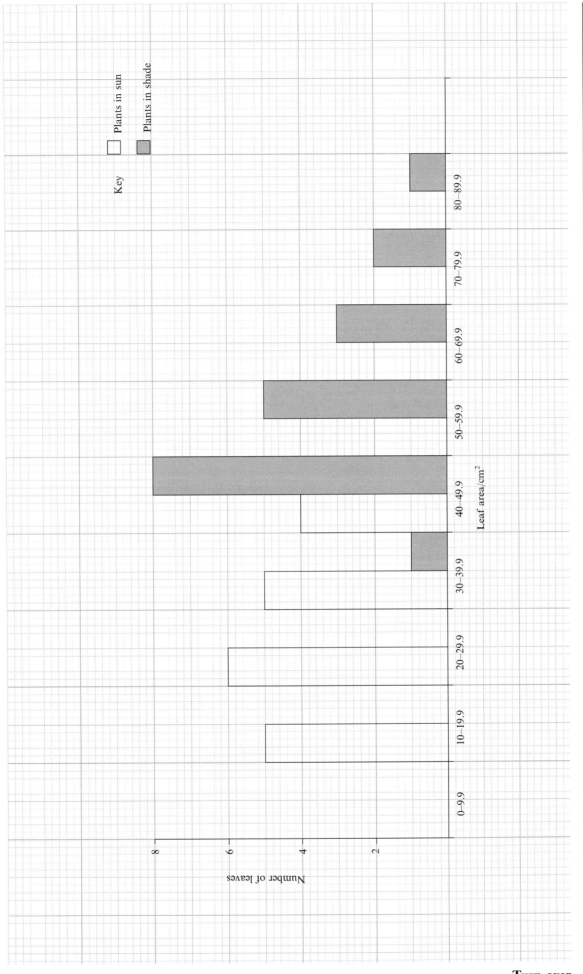

6. Read through the following passage referring to events in the kidney, and complete the passage by writing the most appropriate word or words in the spaces.

High blood pressure in the .... forces water and other small molecules into the Bowmans capsule. This is known as ...., and is made more efficient by .... cells, which have major and minor processes. In the .... convoluted tubule approximately 90% of the .... and all of the amino acids and sugars are reabsorbed. Cells lining this part of the nephron have a .... border and large numbers of .... to provide energy. The loop of Henle acts as a .... multiplier to produce .... urine. It does this by causing a build-up of .... in the medulla of the kidney, causing water to move out of the tubule by ....

ADH is released from the .... pituitary gland. It .... the permeability of the walls of the .... convoluted tubule, so that .... water is reabsorbed into the blood.

(15 marks)

(Total 15 marks)

7. The diagram below shows a section through a *Helianthus* stem.

(a) Comment on the distribution of the vascular bundles in the stem.

....

....

....

(2 marks)

(b) (i) Label with a C **one** area where you would expect to find collenchyma cells.

(1 mark)

(ii) State **one** way that collenchyma cells are adapted to provide support.

....

(1 mark)

(c) (i) Label with an S **one** area where you would expect to find sclerenchyma cells.

(1 mark)

(ii) State **one** way that they are adapted to provide support.

....

(1 mark)

(d) (i) Label with an X **one** area where you would expect to find xylem tissue.

(1 mark)

(ii) State **two** functions of xylem tissue in plants.

....

....

(2 marks)

(e) (i) Label with a P **one** area where you would expect to find phloem tissue.

(1 mark)

(ii) Phloem tissue consists of several different cell types ie. it is a compound tissue. Name **two** types of cell found in phloem tissue.

.... and ....

(2 marks)

(Total 12 marks)

8. The diagram below shows a plant cell placed in a salt solution with a water potential (ψ) of $-400\,\text{kPa}$.

The pressure potential of the cell is $400\,\text{kPa}$, and the solute potential is $-500\,\text{kPa}$.

$\psi_{\text{salt solution}} = -400\,\text{kPa}$

$\psi_p = 400\,\text{kPa}$
$\psi_s = -500\,\text{kPa}$

(a) Calculate the water potential of this cell. **(1 mark)**

(b) (i) Will water move into or out of this cell? ....

(1 mark)

(ii) Give a reason for your answer to (b)(i). ....

....

(1 mark)

(c) A student carried out an investigation into osmosis in potato tubers. She cut cylinders of potato using a cork borer, then cut each of these into discs. The mass of each disc was recorded, then they were placed in a range of salt solutions for 6 hours. Each disc was then blotted dry, and reweighed. The table overleaf shows the data collected.

Turn over

Molarity of salt solution (M)	Original mass (g)	Final mass (g)	Change in mass (g)	Percentage change in mass
0	2.15	2.46	+0.31	+14.42
0.1	2.02	2.06	+0.04	+1.98
0.2	2.04	1.84	−0.20	−9.80
0.3	2.12	1.76	−0.36	−16.98
0.4	2.16	1.47	−0.69	−31.94
0.5	2.05	1.29	−0.76	−37.07
0.6	2.14	1.29	−0.85	−39.72
0.7	2.07	1.15	−0.92	−44.44
0.8	2.19	1.28	−0.91	−41.55
0.9	2.15	1.33	−0.82	−38.14
1.0	2.13	1.20	−0.93	−43.66

(i) Using the grid below, plot a graph to show percentage change in mass at each concentration of salt solution.

(4 marks)

    (ii) What is the concentration of cell sap inside the potato cells?

....

(1 mark)

    (iii) Explain the shape of the graph between concentrations of 0.6 M and 0.9 M.

....

....

(1 mark)

(d) State **two** precautions the student should take to ensure that this is a fair test.

....

....

(2 marks)

(e) Explain the importance of turgor in plant cells.

....

....

....

(2 marks)

(Total 13 marks)

9. In rabbits, coat colour is determined by a single gene which has four alleles:

C^F Full coat colour
C^{CH} Chinchilla
C^H Himalayan
C^A Albino

There is a dominance series, as follows:

C^F is dominant to C^{CH}, C^H, and C^A
C^{CH} is dominant to C^H and C^A
C^H is dominant to C^A

This gene is not sex-linked.

A heterozygous full colour rabbit was mated with a heterozygous chinchilla rabbit, and the offspring were in the ratio 50% full colour, 25% chinchilla and 25% himalayan.

One of the himalayan rabbits was then backcrossed, and the offspring of this cross were 50% himalayan rabbits and 50% albino rabbits.

(a) What do you understand by the term 'backcross'?

....

(1 mark)

(b) (i) What was the genotype of the himalayan rabbit mated in the backcross?

....

(1 mark)

(ii) Show the evidence for your answer to (b)(i).　　　　　**(2 marks)**

(c) Deduce the genotypes of the full colour and chinchilla rabbits mated in the first cross, showing the evidence for your answer.　　　　　**(3 marks)**

(d) Himalayan rabbits are born pure white, but develop black markings on their ears, nose, paws and tail if they are exposed to cold. These markings are then permanent. Suggest how this change in colour could be caused.

....

....

(2 marks)

(Total 9 marks)

10. **Answer the following question on separate ruled paper.**

Give an account of the different types of disaccharides found in living organisms.

(Total 10 marks)

Total marks for paper: 100

Longman
Examination Board

General Certificate of Education

A-level Biology

Paper 3

Time: 2 hours

Number	Mark
1.	
2.	
3.	
4.	
5.	
6.	
7.	
8.	
9.	
Total	

Instructions

■ Attempt ALL the questions.

■ Answer Questions 1–8 in the spaces provided on this exam paper.
Answer Question 9 on separate ruled paper.

■ Show all stages in any calculation, and state the units.

■ Where diagrams are required, draw and label them clearly.

Information for candidates

■ The marks available are shown in brackets after each question or part-question.

■ This exam paper has 9 questions.

■ You are allowed 2 hours for this paper.

■ The maximum mark for this paper is 100.

---

1. The diagram below shows a mitochondrion.

5 μm

Leave margin blank

   (a) Calculate the magnification of this diagram, showing all working.   **(2 marks)**

   (b) In the spaces provided on the diagram, name the parts labelled **A**, **B** and **C**.

   (3 marks)

   Turn over

   25

(c) Name the parts of the mitochondrion in which the following occur:

(i) the Krebs cycle .... **(1 mark)**

(ii) the electron transport chain. .... **(1 mark)**

(d) What do you understand by the term 'oxidative phosphorylation'?

....

....

(2 marks)

(e) One mole of glucose contains 2880 kJ of energy.

During aerobic respiration, one mole of glucose is converted to 38 moles of ATP.

Each mole of ATP is equivalent to 30.6 kJ of energy.

Use the information above to calculate the efficiency of the respiration reaction, expressing your answer as a percentage.

(2 marks)

(Total 11 marks)

2. Sickle cell anaemia is a serious genetic disease caused by a single faulty allele.

Sufferers are homozygous for this allele (Hb^SHb^S), which causes faulty haemoglobin.

Unaffected individuals have the alleles (Hb^AHb^A), and have normal haemoglobin.

(a) What type of mutation causes sickle cell anaemia?

....

(1 mark)

(b) 2% of babies born in a particular population suffer from sickle cell anaemia.

Use the Hardy–Weinberg equation to calculate the percentage of this population who have completely normal haemoglobin.

$p^2 + 2pq + q^2 = 1$

(2 marks)

(c) Which of the following genotypes gives the individual increased resistance to malaria? Underline the correct answer.

Hb^AHb^A Hb^AHb^S Hb^SHb^S **(1 mark)**

(d)   Suggest why individuals with this genotype may be resistant to the malarial parasite.

....

....

(1 mark)

(Total 5 marks)

3.   This question refers to reproduction in plants.

(a)   Explain why cross-pollination is beneficial to plants.

....

(1 mark)

(b)   State **two** ways that plants may favour cross-pollination over self-pollination.

....

....

(2 marks)

(c)   The diagram below shows a plant ovule just prior to fertilisation.

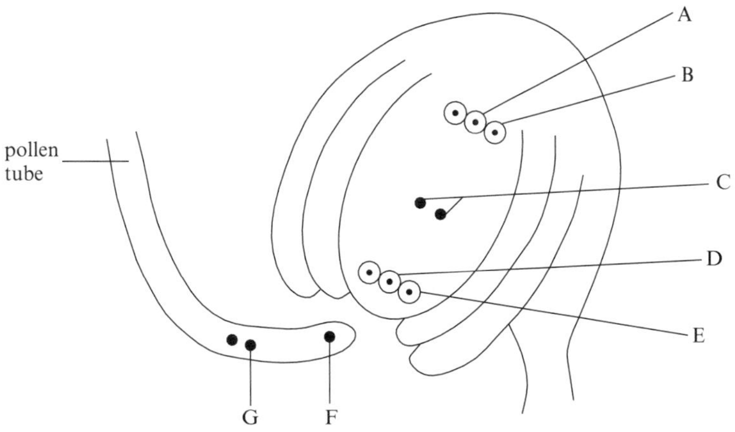

Which letter(s) represent:

(i)   the female gamete?   ....   **(1 mark)**

(ii)   the male gamete?   ....   **(1 mark)**

(iii)   the parts which will fuse to form the endosperm nucleus?   ....

(1 mark)

(d)   What is the role of the part labelled **F**?

....

(1 mark)

Turn over

(e) The graph below shows the changes in mass which occur when a dormant seed becomes active and starts to germinate.

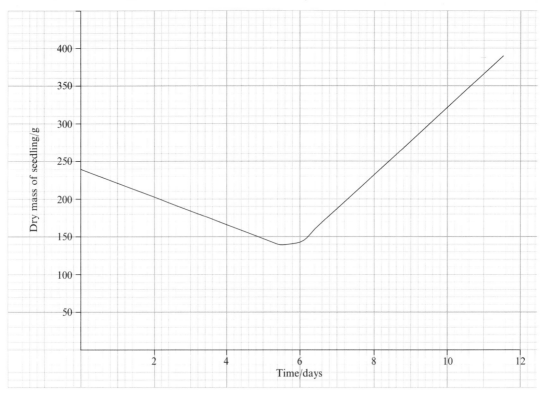

(i) Name **one** hormone associated with breaking of dormancy.

....

(1 mark)

(ii) Why does the seed lose mass in the first 6 days?

....

....

(1 mark)

(iii) Why does it gain mass after this?

....

....

(1 mark)

(Total 10 marks)

4. The graph below shows the changes in the numbers of predators and their prey over time.

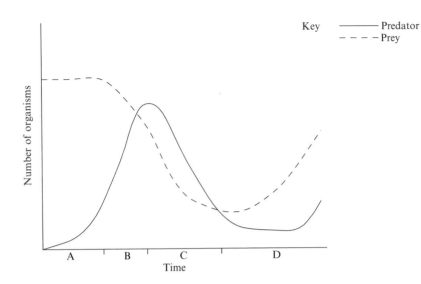

Key   ———— Predator
             – – – – Prey

(a) Explain the changes in the numbers of predators and prey in the following stages of the graph:

B ....

....

....

C ....

....

....

D ....

....

....

(6 marks)

(b) Predators are sometimes used as agents of biological control.

  (i) Explain what this means.

    ....

    ....

    (1 mark)

  (ii) Give **two** advantages of using biological control methods.

    ....

    ....

    (2 marks)

Turn over

Leave margin blank

(iii) Give **two** advantages of using chemical methods to control pests.

....

....

(2 marks)

(iv) What problems could arise if inappropriate predators were used as agents of biological control?

....

....

(1 mark)

(Total 12 marks)

5. *Fucus vesiculosus* is a type of alga (Phaeophyta) found mainly on the middle shore area of rocky shores.

A student collected data to test the hypothesis,

'There is no significant difference between the lengths of *F. vesiculosus* at sample sites A and B.'

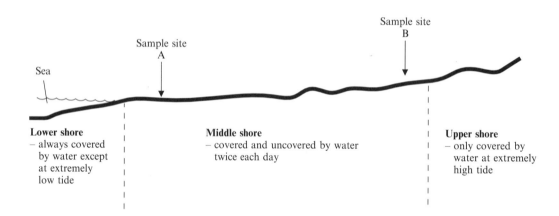

Twenty seaweeds were chosen at random in sample site A, and their length was recorded. This procedure was repeated at sample site B. The table below shows the results obtained.

	Sample site A					Sample site B				
Length of seaweed (cm)	68	49	56	50	74	17	45	24	30	51
	73	62	53	62	49	31	40	46	29	30
	50	55	48	53	56	32	37	32	39	44
	64	56	64	54	63	30	31	46	29	48
Mean length of seaweed (cm)	57.95									

(a) Calculate the mean length of seaweeds at sample site B, and add this to the table.

(1 mark)

(b) Sort the data into groups depending on length, and complete the table below.

Seaweed length (cm)	Tally for site A	Total for site A	Tally for site B	Total for site B
10–19.9				
20–29.9				
30–39.9				
40–49.9				
50–59.9				
60–69.9				
70–79.9				

(3 marks)

(c) Plot histograms of this data on the axes provided below.

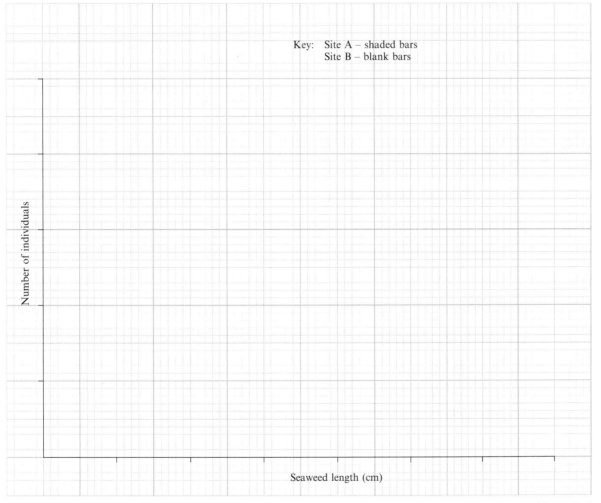

Key: Site A – shaded bars
Site B – blank bars

Number of individuals

Seaweed length (cm)

(4 marks)

(d) In order to find out if the data fits the hypothesis we can carry out a *t*-test.

Use the equation to calculate the value of *t*. **(1 mark)**

$$t = \frac{X_A - X_B}{\sqrt{\dfrac{S_A^2}{n_A} + \dfrac{S_B^2}{n_B}}}$$

X_A = mean of sample site A
X_B = mean of sample site B
S_A^2 = variance of sample A = 59.35
S_B^2 = variance of sample B = 77.19
n_A = number of individuals at site A
n_B = number of individuals at site B

Turn over

(e) (i) At $p = 0.05$, the published value for t is 2.02. Will you accept or reject your hypothesis? .... **(1 mark)**

(ii) Explain the meaning of $p = 0.05$.

....

....

....

(2 marks)

(Total 12 marks)

6. The diagram below shows some of the reactions occurring inside erythrocytes and plasma in capillaries.

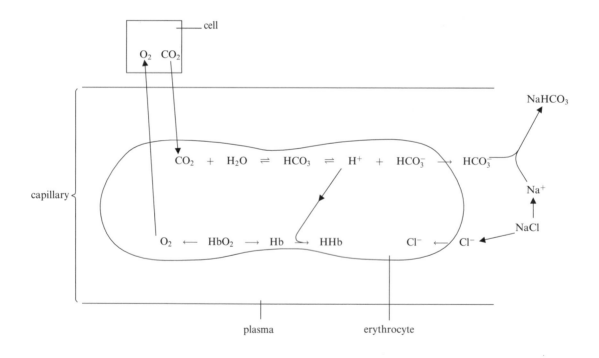

(a) (i) Would this set of reactions be occurring in lung capillaries or muscle capillaries?

....

(1 mark)

(ii) Give a reason for your answer to (a)(i).

....

....

(1 mark)

(b) From the diagram, name:

(i) a conjugated protein with a quaternary structure ....
(1 mark)

(ii) the form in which most carbon dioxide is carried in the blood

....
(1 mark)

(iii) a substance acting as a buffer ....
(1 mark)

(iv) the reaction catalysed by the enzyme carbonic anhydrase.

....
(1 mark)

(c) Why do chloride ions move from the plasma into the erythrocyte?

....

....
(1 mark)

(d) Name **two** substances not shown in the diagram above which would be dissolved in the plasma.

.... and .... **(2 marks)**

(e) Explain how carbon monoxide would affect the amount of oxygen carried by the erythrocyte.

....

....

....
(2 marks)

(Total 11 marks)

7. The area of the foreshore where sand-dunes arise from the beach is unsuitable for the growth of most plants. As sand-dunes increase in age, they will support a changing community of plants: first the pioneer species will colonise the dunes, and successive seral stages will follow.

A survey into sand-dune succession was carried out by a group of students, using the belt transect method. The species present and the dominant species at each sampling point were recorded. Samples of sand were collected from each quadrat and analysed in the laboratory, for percentage moisture and percentage humus content. The results are shown in the table overleaf.

Turn over

Distance along transect (m)	Number of species present	Names of species present (dominant species in bold type)	% water in sample	% humus in sample
0	1	marram grass	0.82	0
5	1	marram grass	1.33	2.6
10	2	**marram grass** sea holly	2.81	3.4
15	2	**marram grass** rest harrow*	3.00	3.9
20	3	marram grass **rest harrow*** kidney vetch*	4.87	5.3
25	3	**fescue** rest harrow* sea bindweed	5.90	6.5
30	4	kidney vetch* **fescue** rest harrow* lesser hawkbit	6.63	8.7
35	5	fescue **rest harrow*** common cats ear lesser hawkbit plantain	7.37	10.2
40	4	fescue bird's foot trefoil* **kidney vetch*** common cats ear	7.52	10.9
45	6	fescue dewberry **common cats ear** kidney vetch* plantain	8.11	12.4
50	6	fescue creeping willow **dewberry** plantain white clover* red clover*	8.73	12.1

(a) (i) Suggest **two** reasons why sand-dunes close to the beach might be unsuitable for the growth of most plants.

....

....

(2 marks)

(ii) From the data in the table, identify the pioneer species which **first** colonises bare sand.

....

(1 mark)

(iii) Suggest and explain **one** adaptation that this species may possess to make it successful in this habitat.

....

....

....

....

(2 marks)

(b) (i) State the relationship between the amount of humus and the age of the sand-dune.

....

....

(1 mark)

(ii) Explain how this change occurs.

....

....

....

....

(2 marks)

(iii) Explain why percentage humus levels and percentage moisture levels seem to be linked.

....

....

(1 mark)

(iv) Why does humus increase soil fertility?

....

....

....

(1 mark)

(c) (i) Species marked with an asterisk (*) are leguminous plants. Why might these plants have a selective advantage in the sand-dune habitat?

....

....

....

....

....

(4 marks)

Turn over

(ii) Suggest **two** types of plants not listed so far which you might see if succession continued and the site reached a climax community.

.... and .... **(2 marks)**

(Total 16 marks)

8. *Chlorophytum* (spider plant) has variegated leaves.

A student carried out an investigation to compare the numbers of stomata in different areas of *Chlorophytum* leaves. Nail varnish peels of different regions of the leaf were made, and viewed under the microscope. The number of stomata per field of view was then recorded.

(a)   (i)   **How**, and **why**, would you expect the number of stomata to vary in the green area of the lower leaf surface, compared to the white area of the lower leaf surface?

....

....

....

(2 marks)

(ii)   **How**, and **why**, would you expect the number of stomata to vary if you compared the green area of the upper surface with the green area of the lower surface?

....

....

....

(2 marks)

(iii)   Suggest **one** way that the student could increase the validity of his results.

....

(1 mark)

(b)   The diagram below shows a potometer.

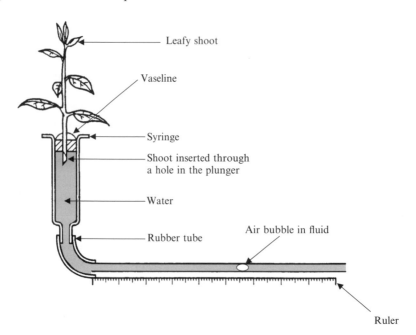

Leafy shoot

Vaseline

Syringe

Shoot inserted through a hole in the plunger

Water

Rubber tube

Air bubble in fluid

Ruler

(i) State briefly how you could use this apparatus to measure transpiration rate.

....

....

(1 mark)

(ii) The graph below shows how transpiration rate varied over a 24-hour period.

Explain the trends shown in this graph.

....

....

....

....

....

(2 marks)

(c) (i) Apart from the time of day, state **one** other factor which affects transpiration rate.

....

(1 mark)

(ii) Explain how you could investigate this, using the potometer.

....

....

....

....

(3 marks)

Turn over

(d)   Why is transpiration beneficial to plants, even though they lose a lot of water this way?

....

....

(1 mark)

(Total 13 marks)

9.   **Answer the following question on separate ruled paper.**

Give an account of gas exchange in insects.

(Total 10 marks)

Total marks for paper: 100

Longman Examination Board

General Certificate of Education

A-level Biology

Paper 4

Time: 2 hours

Number	Mark
1.	
2.	
3.	
4.	
5.	
6.	
Total	

Instructions

■ This paper has six questions. You should answer any **three** questions.

■ Write your answers on separate ruled paper.

Information for candidates

■ Each question is worth 20 marks.

■ You are allowed 2 hours for this paper.

■ The maximum mark for this paper is 60.

All questions are worth 20 marks.

Answer **three** questions.

Leave margin blank

You are reminded of the necessity for good English (grammar, punctuation and spelling) and orderly presentation in your answers.

1. Give an account of the importance of water to living organisms.

2. Describe the distribution and roles of membranes in cells.

3. Compare reproduction in humans and flowering plants.

4. Discuss the significance of micro-organisms to humans.

5. Give an account of the structure and importance of nucleic acids.

6. How has the work of each of the following scientists increased our understanding of biology?

   (a) Mendel

   (b) Darwin

   (c) Watson and Crick

   (d) Jacob and Monod

Total marks for paper: 60

Solutions to practice exam papers

General instructions

Try to follow the mark scheme as closely as possible, but remember that it is unlikely that you will use exactly the same phrases in your answer.

;   Indicates information needed to gain one mark.

/   Indicates acceptable alternative answers – you cannot gain 2 marks for 2 of these.

( )   Terms in brackets are not needed to gain marks – they are included to show the steps needed to reach the final answer.

eq   Any sensible answer will gain a mark (examples of suitable answers will be given).

Solutions to Paper 1

1. You gain one mark for each correct line, so do not leave any gaps.

     ✓   ✓   ✗   ;

     ✓   ✓   ✓   ;

     ✓   ✓   ✗   ;

     ✗   ✓   ✗   ;

     ✓   ✗   ✗   ;

     ✗   ✗   ✗   ;

Total 6 marks

> **TIP**
>
> Make sure that your ticks and crosses are absolutely clear: you will gain no credit for answers which look like this ✗.

2. (a)   A   Reverse transcriptase;

         B   DNA polymerase;

         C   Restriction endonuclease;

         D   DNA ligase;   **4 marks**

    (b)   Anterior pituitary gland;   **1 mark**

    (c)   Temperature; use a cooling jacket (to maintain optimum temperature);

         pH; add buffer or alkali (to maintain optimum pH);

         oxygen; pump in sterile air / oxygen (needed for respiration);

         nutrients; add controlled amounts of a suitable carbon source, nitrogen source, etc.;

                                                       3 marks

> **TIP**
>
> The question asks for HOW the factors are controlled, not WHY. Make sure you include the correct information in your answer.

Total 8 marks

3. Thylakoids / grana; stroma; chlorophyll / photosystems I and II; higher; ADP; ATP; water; oxygen; NADP; $NADPH^+$ / $NADPH_2$; ribulose bisphosphate; ribulose bisphosphate carboxylase (Rubisco);

Total 12 marks

4. LH (luteinising hormone); glucagon; acetylcholine; progesterone; oxytocin; noradrenaline.

6 marks

Total 6 marks

5. (a) (i) Rapid growth at first;
for first 8 days;
then curve levels out / population remains constant;
plateau at 100 units **2 marks**

(ii) Density dependent;
population remains stable / does not crash;
not reduced by external factors; **2 marks**

(b) *P. caudatum* increases rapidly at first / for first 8 days;
then stabilises at 25 units;
then decreases;
to zero by day 20;

P. aurelia population increases very slowly for the first 2 days;
then increases rapidly up to 12 days;
then levels off; **4 marks**

(c) (i) The population increases rapidly because they are able to obtain the food they need /
the reproductive rate is very fast. **1 mark**

(ii) The population decreases because they cannot compete with *P. aurelia* for food;
1 mark

> **TIP**
>
> In part (b) you are being asked to describe the graph; in part (c) you are asked to explain it.
> **Describe** means you should write about the shape of the graph, using figures taken from it.
> **Explain** means you should write about the underlying biological reasons for it, i.e. relate your
> theory knowledge to this data.

(d) When grown alone, *P. aurelia* reaches maximum population size early, and stabilises;
when grown together, *P. aurelia* takes longer to reach maximum population size / only
reaches maximum population size when *P. caudatum* is extinct;
maximum population size is the same in both experiments; **2 marks**

(e) Two species in the same community cannot occupy the same ecological niche. **1 mark**

Total 13 marks

6. (a) Number of cells in chamber $= 27$;

volume of liquid in chamber $= 2.5 \times 10^{-4}$ ml

in 1 ml there will be $\dfrac{27}{2.5 \times 10^{-4}}$;

$= 1.08 \times 10^5$ cells per ml; **3 marks**

> **TIP**
>
> Calculations like this are very common. Make sure you understand dilution factors, and that you
> are confident working with the large numbers involved. Always mark the counting grid to show
> evidence of a systematic counting method (e.g. by crossing through cells as you count them), as
> marks may be given for this.

(b) This sample was diluted $\dfrac{1}{100}$ i.e. 10^{-2}

undiluted sample contains $1.08 \times 10^5 \times 100 = 1.08 \times 10^7$ cells per ml; **1 mark**

(c)   Numbers of cells are too large to fit easily on the graph scale;   **1 mark**

(d)   (lag phase / slow growth) Cells are synthesising new enzymes / becoming used to new conditions for growth;

(Exponential phase / log phase) factors are at an optimum for rapid cell division;

(Stationary phase / plateau) lack of nutrients / build-up of waste products causes cell division to slow down;   **3 marks**

(e)   (i)

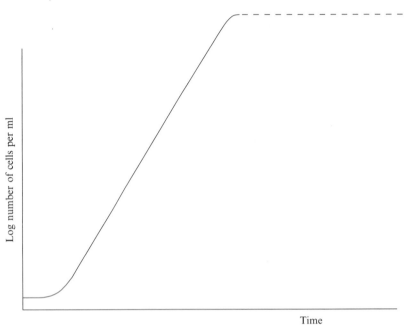

         1 mark

(ii)   Cell division stops;

This counting method does not distinguish between live and dead cells / as cells die, the number of cells remains constant;   **2 marks**

         Total 11 marks

7.   (a)   (i) G   (ii) B / C / F   (iii) E / B   (iv) C   (v) B / C / F   **5 marks**

(b)   Activated tRNA / tRNA with an amino acid attached;
tRNA lines up against mRNA;
Anti-codon of tRNA bonds to codon of mRNA;
ribosome holds tRNA and mRNA in place;
while peptide bonds form between amino acids;
tRNA is then released;
ribosome moves along mRNA;
correct reference to start / stop codons.

         4 marks

         Total 9 marks

8.   (a)   (Each cycle takes 0.8 s)

Number of beats per minute $= \dfrac{60}{0.8} = 75$ **1 mark**

(b)   ventricles contract;
movement of blood / increased blood pressure closes atrio-ventricular valves;   **2 marks**

(c)   (i) 0.4 s;   **1 mark**

(ii) Ventricular and aortic pressures become different as the valve closes;   **1 mark**

(d) The walls of the aorta contain elastic fibres so recoil occurs; (this maintains pressure)

1 mark

(e) Medulla oblongata;   **1 mark**

(f) (i)

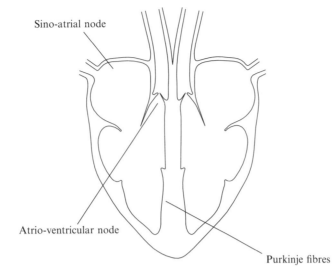

Sino-atrial node

Atrio-ventricular node

Purkinje fibres   **3 marks**

TIP

Clear labelling is vital to gain full marks. It is better to use label lines than to write on the diagram.

(ii) **SA node** – initiates contraction / acts as a pacemaker;
AV node – carries electrical impulse from the atria to the ventricles (a non-conducting septum prevents impulses passing freely);
Purkinje fibres – carry impulse to the apex of the heart / make the ventricles contract from the bottom up / delay the start of contraction of the ventricles;

3 marks

Total 13 marks

9. (a) Mean achromatic time: 19, 18.5, 17.5, 15, 13.5, 12, 11.5, 10, 9.5;

Rate of reaction: 0.053, 0.054, 0.057, 0.067, 0.074, 0.083, 0.087, 0.1, 0.11;   **2 marks**

TIP

If you are not sure how to calculate rate of reaction, look carefully at the first line which has been done for you.

$$\text{Rate} = \frac{1}{\text{time taken for reaction to occur}}$$

(b) Marks are given for choosing a sensible scale; for labelling the axes correctly; for neat, clear points, correctly plotted; for line of best fit / smooth curve;

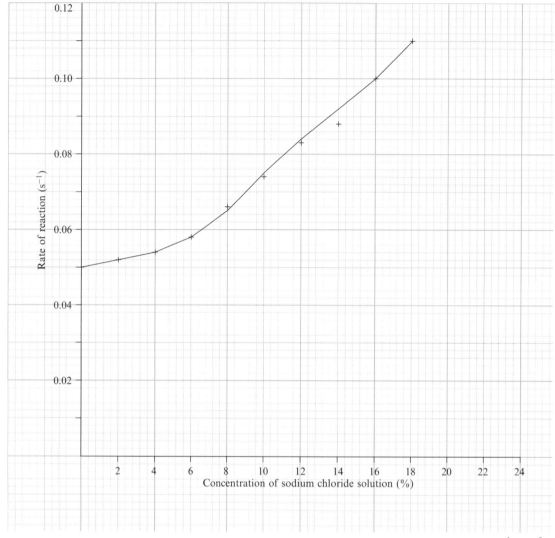

4 marks

(c) As the concentration of sodium chloride increases, the rate of reaction increases / amylase works better; **1 mark**

(d) Sodium chloride is an enzyme activator;
it is needed for optimum efficiency; **2 marks**

(e) Amylase may be denatured;
hydrogen / ionic bonds are broken, so the active site changes shape; **1 mark**

(f) Carry the experiment out in a waterbath, to maintain optimum temperature;
test samples every 30 s;
test a wider range of concentrations;
carry out more repeats; **2 marks**

TIP

When you are asked for experimental details, marks are usually available for
 ■ controlling the variable being investigated
 ■ information on the range of variables to be tested
 ■ controlling the other variables (fair test)
 ■ a reliable way of collecting results
 ■ taking repeat readings and averaging.

Total 12 marks

10. The functional unit of the muscle is the sarcomere;
each sarcomere has light /isotropic bands and dark / anisotropic bands;
two types of protein are found in each myofibril;
actin consists of thin filaments;
myosin consists of thick filaments;
actin and myosin filaments overlap / diagram to show overlapping;
correct reference to M-line / H-zone / Z-line;
tropomyosin and troponin are also present;
sliding filament theory of contraction / fibres do not change in length, they slide past each other;
myosin head attaches to actin;
myosin head changes position, pulling actin towards the M-line;
myosin head detaches from the actin filament;
ATP is needed to return the myosin head to its original position;
this is like a ratchet mechanism;
calcium ions are needed for contraction to occur;
calcium ions bind to troponin;
diagram to show final appearance of the sarcomere / description of the changes
e.g. I-band shortens, etc.

Total 10 marks

(Total marks for paper: 100)

Solutions to Paper 2

1.   × ✓ × ;

   ✓ ✓ × ;

   ✓ ✓ × ;

   × ✓ × ;

   ✓ ✓ × ;

   ✓ ✓ ✓ ;

Total 6 marks

2. (a) (i) C   **1 mark**

   (ii) F   **1 mark**

   (iii) I   **1 mark**

   (iv) F   **1 mark**

 (b) (i) False; (there is at first, but it levels off)   **1 mark**

   (ii) False; (temperature is also a limiting factor)   **1 mark**

   (iii) False; (it is not possible to deduce this from the data given) **1 mark**

   (iv) True; (compare curves 2 and 3)   **1 mark**

TIP

You are told to **use only the information given in the graph**. This means that you must consider the graph very carefully: for example, statement 3 is true, but it is not possible to deduce this from the graph, so it must be marked false.

(c)

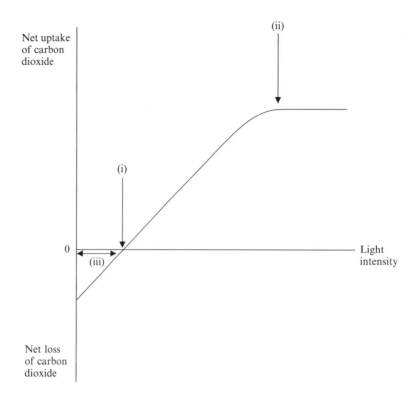

3 marks

Total 11 marks

3. (a) (i) Meiosis   **1 mark**

    (ii) Prophase I;   **1 mark**

    (iii) Because synapsis has occurred;
         bivalents have formed;
         crossing over has occurred / chiasmata are present;   **2 marks**

    (iv) Crossing over;   **1 mark**

    (v) This separates linked genes;
       chromosomes now contain different information;
       gametes contain different combinations of genes;   **2 marks**

(b) (i) A – genetically identical to sheep X;   **1 mark**

    (ii) A – true, because it has identical DNA to sheep X;
       B and C – false, because phenotype depends on environmental factors as well as
       genetic factors;
       D – false;   **2 marks**

Total 10 marks

4. (a) <u>In Birmingham, melanic moths survive longer than light moths.</u>   **1 mark**

(b) Melanic moths are mutant / arise as a result of a mutation;
     in industrial areas, melanic moths are better camouflaged;
     they are likely to live longer (they have a selective advantage);
     they will breed, and produce offspring like themselves;
     the proportion of melanic moths in the population will rise;   **4 marks**

(c) *Polymorphism* = a range of clearly different, genetically determined phenotypes is present
     in the population;
     e.g. blood group in humans / banding pattern in land snails (*Cepaea*);   **2 marks**

Total 7 marks

5. (a) Draw round a leaf on squared paper; **1 mark**

    (b) (i) Leaves in shade grow larger to absorb more of the available light; **2 marks**

        (ii) Height – shade plants grow taller / amount of chlorophyll – shade plants have more; **1 mark**

    (c) *t*-test; **1 mark**

    (d) 95% of the sample is covered by the range 27.55 ± 2 SD;

$$= 27.55 + 19.24 = 46.79$$

$$= 27.55 - 19.24 = 8.31$$

Range $= 8.31$ to 46.79; **2 marks**

Total 7 marks

6. Glomerulus; ultrafiltration; podocyte; proximal; water; brush / microvilli; mitochondria; counter-current; hypertonic; sodium chloride / salt; osmosis; posterior; increases; distal; more. **Total 15 marks**

7. (a) Located close to the outside of the stem;

to withstand bending forces / to make the stem stronger; **2 marks**

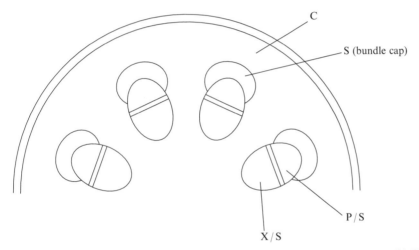

(b)(i) Label **C** correct **1 mark**
(c)(i) Label **S** correct **1 mark**
(d)(i) Label **X** correct **1 mark**
(e)(i) Label **P** correct **1 mark**

    (b) (ii) Extra cellulose thickening in the corners of the cell wall; **1 mark**

    (c) (ii) Cell walls thickened with lignin; **1 mark**

    (d) (ii) Carry water;
           carry minerals;
           support the plant; **2 marks**

    (e) (ii) Companion cell; sieve tube element; parenchyma; sclereids; fibres. **2 marks**

> **TIP**
>
> Clear labelling is vital here. Students often avoid learning plant anatomy, but the questions set are usually very straightforward.

Total 12 marks

8. (a) $(\psi = \psi_S + \psi_P)$

$(= -500 + 400)$

$= -100\,\text{kPa};$ **1 mark**

TIP

Make sure that you know the equation for calculating water potential, and that you understand the direction of water movement.

(b) (i) Out of the cell; **1 mark**

(ii) Water always moves to an area of lower water potential; (i.e. from $-100\,\text{kPa}$ to $-400\,\text{kPa}$) **1 mark**

(c) (i) Marks are given for suitable scale (including increase and decrease in mass); for labelling axes; for neat, clear points, plotted correctly; for smooth curve;

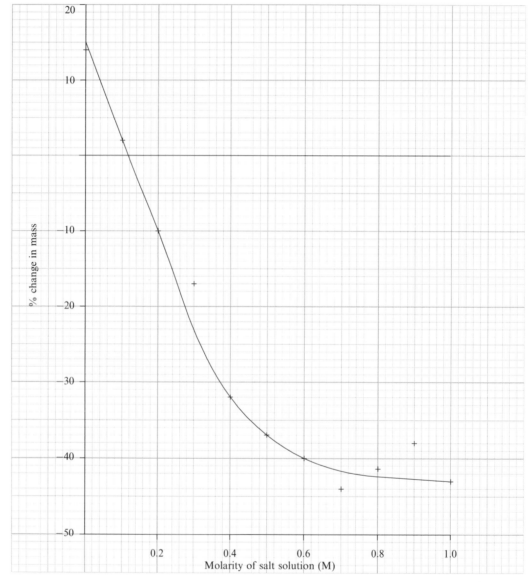

4 marks

(ii) 0.11 M; (read off the graph at the point at which there is zero change in mass) **1 mark**

(iii) Cell is fully turgid so it cannot absorb any more water; **1 mark**

(d) Place several discs in each solution to take an average;
take all discs from the same potato;
all discs must be the same size / have the same surface area; **2 marks**

(e) Cells swell and pack together tightly;
provides support in herbaceous (non-woody) plants;
particularly important in parenchyma tissue; **2 marks**

Total 13 marks

9. (a) Cross with the homozygous recessive ($C^A C^A$); **1 mark**

(b) (i) Himalayan rabbit $= C^H C^A$; **1 mark**

(ii) (parents C^H C^A C^A C^A)

gametes 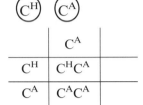;

	C^A
C^H	$C^H C^A$
C^A	$C^A C^A$

cross correct (1 mark)

2 marks

(c) There are two possible genotypes for the parents; either is acceptable.

$$C^F C^A \quad \text{and} \quad C^{CH} C^H \qquad \text{or} \qquad C^F C^H \quad \text{and} \quad C^{CH} C^A;$$

gametes   ;

	C^{CH}	C^H
C^F	$C^F C^{CH}$	$C^F C^H$
C^A	$C^{CH} C^A$	$C^H C^A$

	C^{CH}	C^A
C^F	$C^F C^{CH}$	$C^F C^A$
C^H	$C^{CH} C^H$	$C^H C^A$

cross correct (1 mark) **3 marks**

> **TIP**
>
> You will have to use all of the information you are given, and work through questions 9(b) and (c) in a logical way to gain full marks. Try out your answer in rough if necessary, but make sure that you include all the steps of your working in your final answer. Two possible solutions are given.

(d) The gene codes for a temperature sensitive enzyme, i.e. it is only activated at low temperatures; these parts of the body are coldest; **2 marks**

Total 9 marks

10. Disaccharides are formed from two monosaccharide molecules;
by a condensation reaction / water is also a product;
they are joined by a glycosidic bond;
one mark for a correct diagram showing the bond;
general formula is $C_{12}H_{22}O_{11}$;
one mark for naming at least two disaccharides (maltose, lactose or sucrose)
maltose contains glucose and glucose;
it is an intermediate product in the breakdown of starch;
it is common in germinating seeds / in animal digestive systems;
it is a reducing sugar;
lactose contains glucose and galactose;
it is found in milk;
it is a reducing sugar;
sucrose contains glucose and fructose;
it is the main form in which carbohydrate is transported in plants;
it is found in large quantities in sugar cane and sugar beet;
it is a non-reducing sugar. **Total 10 marks**

(Total marks for paper: 100)

Solutions to Paper 3

1. (a) $\text{Magnification} = \dfrac{\text{drawing size}}{\text{actual size}} = \dfrac{8.5\,\text{cm}}{5\,\mu\text{m}}$;

$$= \dfrac{8.5 \times 10^{-2}\,\text{m}}{5 \times 10^{-6}\,\text{m}}$$

$$= 1.70 \times 10^4 / 17\,000 \text{ times; accept } 1.68 \text{ to } 1.73 \times 10^4$$

2 marks

> **TIP**
>
> Magnification calculations are very common – make sure you understand how to do them, and that you can calculate the actual size if you are given the magnification.

(b)   A – envelope / outer membrane;
B – matrix;
C – crista / stalked particles;   **3 marks**

(c)   (i)   matrix / B;   **1 mark**

(ii)   stalked particles of inner membrane / C;   **1 mark**

(d)   adding phosphate to ADP / changing ADP to ATP;
in the presence of oxygen;   **2 marks**

(e)   Total available energy $= 2880$ kJ

Actual energy released $= 38 \times 30.6 = 1162.8$ kJ;

$\%$ efficiency $= \dfrac{1162.8}{2880} \times 100 = 40.38\%$;   **2 marks**

Total 11 marks

2. (a)   Point mutation / substitution mutation;   **1 mark**

(b) $p^2 + 2pq + q^2 = 1$
$p + q = 1$

$q^2 = 2\% = 0.02$
$q = 0.14$;

$p = 0.86$ so p^2 (homozygous dominant) $= 0.74$

74% of the population have completely normal haemoglobin;   **2 marks**

(c) $\underline{Hb^A Hb^S}$;   **1 mark**

(d)   Parasites cannot enter red blood cells / parasites cannot feed on faulty haemoglobin;
1 mark

Total 5 marks

3. (a)   Increases variation / reduces inbreeding;   **1 mark**

(b)   Protandry – stamens ripen before stigma;
protogyny – stigma ripens before stamens;
dioecious plants have all male or female flowers, so self-pollination is impossible;
flower structure may make self-pollination unlikely / heterostyly;   **2 marks**

(c)   (i)   D   **1 mark**

(ii)   G   **1 mark**

(iii)   G and C   **1 mark**

(d) Controls the growth of the pollen tube;   **1 mark**

(e) (i) Gibberellins / ethene / cytokinins;   **1 mark**

    (ii) Food stores in the seed (carbohydrate, lipid) are being used as a substrate for respiration;   **1 mark**

    (iii) Seedling begins to photosynthesise;   **1 mark**

Total 10 marks

4. (a) **B** – Numbers of predators are increasing, because there is a lot of food available; numbers of prey are falling, as more are being eaten;

    C – Numbers of predators begin to fall, as food (prey) is in short supply; numbers of prey continue to fall, as they are eaten by predators;

    D – Numbers of prey begin to rise, as less are being eaten; numbers of predators remain low, but then rises as prey increases;   **6 marks**

(b) (i) Predators are used to control the prey numbers, when the prey is considered a pest;   **1 mark**

    (ii) No chemicals are added to the habitat / does not pollute the habitat; selective / only harms the prey species; low cost / does not need to be frequently reapplied;   **2 marks**

    (iii) Will eliminate pests completely; acts very rapidly;   **2 marks**

    (iv) the predators may target other prey species apart from the pest / predators may themselves become pests;   **1 mark**

Total 12 marks

5. (a) 35.5 cm;   **1 mark**

(b) Marks are given for site **A** data correct; site **B** data correct; tallies completed;

Seaweed length (cm)	Tally for site A	Total for site A	Tally for site B	Total for site B
10–19.9			|	1
20–29.9			|||	3
30–39.9			⫻⫻ ||||	9
40–49.9	|||	3	⫻⫻ |	6
50–59.9	⫻⫻ ||||	9	|	1
60–69.9	⫻⫻ |	6		
70–79.9	||	2		

3 marks

(c)   Marks are given for suitable scale; for site **A** data plotted correctly; for site **B** data plotted correctly; for shading correct (as stated in key);

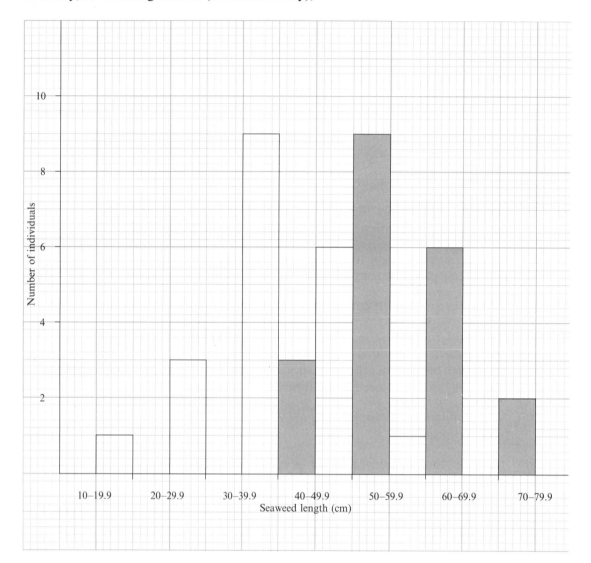

4 marks

(d) $t = \dfrac{57.95 - 35.5}{\sqrt{\dfrac{59.35}{20} + \dfrac{77.19}{20}}}$

  $= \dfrac{22.45}{\sqrt{2.97 + 3.86}} = 8.6$ (One mark for correct answer)   **1 mark**

(e)   (i)   Reject the hypothesis;   **1 mark**

  (ii) $p = 0.05$ means that there is a 5% chance that we are wrong to reject the hypothesis;   **2 marks**

TIP

In statistics, it is usually impossible to be certain that your analysis of results is correct. Biologists normally work at the 5% confidence level ($p = 0.05$): this applies to all of the tests you are likely to meet at A-level.

Total 12 marks

6. (a) (i) Muscle capillaries; **1 mark**

   (ii) Oxygen is diffusing out of the capillary / carbon dioxide is diffusing into the capillary; **1 mark**

   (b) (i) Haemoglobin; **1 mark**

   (ii) Na H CO₃/sodium hydrogen carbonate; **1 mark**

   (iii) Haemoglobin; **1 mark**

   (iv) $H_2O + CO_3 \rightleftharpoons H_2CO_3$; **1 mark**

   (c) To maintain the electrochemical neutrality / to replace the HCO_3^- ions which have diffused out of the cell; **1 mark**

> **TIP**
>
> This is the chloride shift, but you were asked to explain why it occurs, not simply to name it.

   (d) Glucose / amino acids / glycerol / K^+ ions/ Ca^{2+} ions / hormones / **eq** **2 marks**

   (e) The amount of oxygen carried would be reduced;
   haemoglobin binds to carbon monoxide in preference to oxygen / binds to carbon monoxide irreversibly; **2 marks**

   Total 11 marks

7. (a) (i) Low levels of water;
   low levels of minerals / nitrate;
   high levels of salt; **2 marks**

   (ii) Marram grass; **1 mark**

   (iii) Rolled leaves;
   stomata in pits;   } all reduce water loss by transpiration;
   hairs on leaves;
   thick, waxy cuticle;
   long roots; to absorb all available water;
   thrives when covered by loose sand; highly competitive in sand-dune habitat; **2 marks**

> **TIP**
>
> Marram grass is a xerophyte (grows in conditions where water is scarce). Most of the features listed above are typical xerophyte features – you should know these, even if you are not familiar with marram grass.

   (b) (i) As the age of the sand-dune increases, the amount of humus increases; **1 mark**

   (ii) Pioneer species colonise the dune;
   when they die, they decompose, leaving humus;
   humus builds up as generations of plant die and decay;
   gradually the soil becomes suitable for a wider variety of species;
   succession occurs / the habitat passes through seral stages; **2 marks**

   (iii) Humus soaks up water / prevents free drainage of the soil; **1 mark**

   (iv) Humus releases minerals as it decays;
   humus retains water, so it reduces leaching of minerals in rainwater; **1 mark**

   (c) (i) They have root nodules containing nitrogen fixing bacteria;
   they can use nitrogen gas to make nitrates;
   they use this to make proteins / nucleic acids / chlorophyll etc.;
   this gives them an advantage in a habitat where nitrates are scarce; **4 marks**

(ii) Trees / named trees;
shrubs / named shrubs;   **2 marks**

Total 16 marks

8. (a) (i) More stomata in the green area;
green areas will photosynthesise, so need more efficient gas exchange;   **2 marks**

(ii) More stomata on the lower surface;
this reduces water loss by transpiration;   **2 marks**

(iii) Make several peels from each region of the leaf, and calculate mean values;
test several different leaves (of the same species);
count several fields of view;   **1 mark**

(b) (i) Measure how far the bubble moves in a given time;   **1 mark**

> **TIP**
>
> Students often lose marks because they do not mention time when calculating rate.

(ii) Low at night when the stomata are closed;
high during the day, when the stomata are fully open;   **2 marks**

(c) (i) Humidity / air movement / temperature;   **1 mark**

(ii) Humidity – enclose the shoot in a clear polythene bag;
change the humidity by spraying water into the bag / **eq**;
air movement – place the potometer on a moving turntable / near to a fan;
change air movement by regulating the speed of the fan / **eq**;
temperature – place potometer in a clear thermostatically controlled box;
e.g. plant propagator;
keep all other conditions the same (fair test)   **3 marks**

> **TIP**
>
> In part (c)(i) you choose one factor – make sure that you choose one that you can write about in part (c)(ii). Remember to control all other conditions (fair test).

(d) transpiration stream moves water to the upper parts of the plant / moves minerals to the leaves / cools the leaves;   **1 mark**

Total 13 marks

9. The respiratory system in insects is the tracheal system;
blood is not involved in the transport of gases;
exoskeleton is impermeable to gases;
there are holes in the exoskeleton of the thorax and abdomen;
holes are called spiracles;
can be opened and closed by tiny muscles;
lead into a network of tubes called tracheae and tracheoles;
correct description of arrangement of tubes / diagram to show this;
tubes are kept open by rings of chitin;
tracheoles lead directly into cells;
ends of tracheoles contain fluid;
amount of fluid can be varied / there is less when the insect is very active;
ventilation is by changing the body shape;
dorso-ventral muscles contract to flatten the body;
this reduces the volume, so air is forced out;
inspiration is due to elastic recoil / is passive;
chemoreceptors detect levels of carbon dioxide in the body.

Total 10 marks

(Total marks for paper: 100)

Paper 4 – Hints and guidance

> **TIP**
>
> Be prepared to spend at least 5 minutes reading through the questions before you start.

Your first task is to choose the right essays!

■ Structured essays, which give an indication of the subject content required and are divided into sub-sections, are usually easier to write than essays with a very short title.
For example: 'Write about the importance of water to living things.'

■ Look carefully at the wording of the title.
'Give an account of ...' is easier than 'Compare ...' where you will have to include similarities and differences, or 'Discuss ....', where you should give several points of view.

You must plan your essay carefully

■ The plan is the outline of the information to be included.
It covers which information is written, and in which order; this allows you to produce a logical, well-structured essay, addressing all aspects of the title.

Remember, there are lots of ways to write a good essay – the plans and outlines given here cover the information you should include to achieve a high mark, but you must include an appropriate level of detail, and write in a coherent essay style. Refer to pages iv–v for guidance on how essays are marked.

1. **'Give an account of the importance of water to living organisms.'**

   This is quite a difficult essay because you need to include facts from so many different parts of the syllabus to gain high marks. The plan below shows one way you could tackle it.

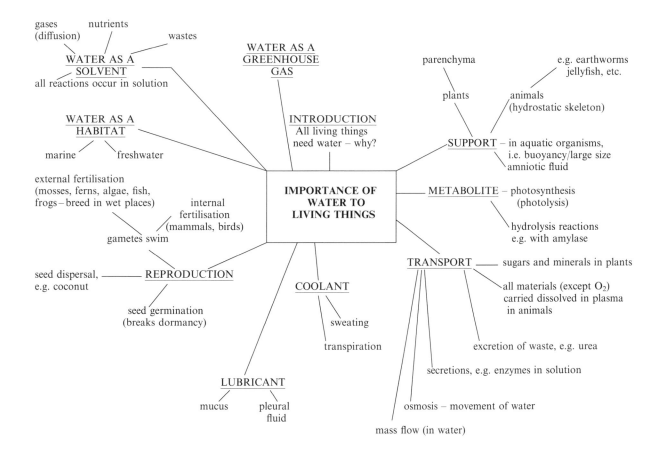

2. **'Describe the distribution and roles of membranes in cells.'**
Again, quite a difficult essay, because you probably have not given much thought to where membranes are and what they do. Do not be tempted to stray from the point and write a lot about the structure of membranes; a brief mention in the introduction is all that is required.

Introduction – components, e.g. phospholipids, proteins, etc.
fluid mosaic model
Distribution – outer membrane
around organelles, e.g. chloroplasts, mitochondria, etc.
within organelles, e.g. cristae of mitochondria, lamellae of chloroplasts, etc.
network within cytoplasm; endoplasmic reticulum, Golgi body, etc.
around vesicles, vacuoles, etc.
Roles – divides the cell into compartments – allows a form of division of labour to occur
controls what enters and leaves the cell (it is semi-permeable)
allows materials to be stored within the cell
allows materials to be packaged prior to secretion
increases the internal surface area of the cell – e.g, for attachment of stalked particles, chlorophyll, etc.
Ending – universal nature of membranes; both prokaryotic and eukaryotic cell have them.

3. **'Compare reproduction in humans and flowering plants.'**
Here you should compare humans with plants all the way through the essay. Do not just write a page about humans, then a page about plants and leave the examiner to work out the comparison!

Introduction – definition of reproduction
many plants can reproduce both asexually and sexually; humans only reproduce sexually
Systems – many plants are hermaphrodite, whereas humans are either male or female
brief description of reproductive systems
Gametes – haploid gametes produced by meiosis in humans, by mitosis in plants (alternation of generations), compare numbers and motility of gametes
Transfer of gametes – pollination in plants / adaptations for pollination, courtship and intercourse in humans
Fertilisation – movement of male gamete to female gamete, fertilisation within the female reproductive system, development of embryo; 'double fertilisation' in plants
Growth of embryo – nutrition from parent, laying down food store in plant, dormancy
Birth / dispersal – brief description of events
Parental care – extended in humans (give examples)
Involvement of hormones
Possibility of self-pollination in plants
Ending – compare the relative numbers of offspring produced
– comment on the most effective strategy.

4. **'Discuss the significance of micro-organisms to humans'**
This is a much more straightforward essay. Make sure that you include ways that microbes are of use to humans as well as the ways that they are harmful. Give named examples wherever possible, and outlines of the processes involved.

Introduction – definition of microbes
brief description of bacteria, fungi, protoctists and viruses
Uses – food production – brewing, baking (yeasts)
yogurt, cheese (lactic acid bacteria)
fermented products, e.g. soy sauce,
single cell protein

pharmaceuticals, e.g, antibiotics, vaccines

agriculture, e.g, silage production

enzymes, e.g. for washing powders, textiles, confectionery, etc.

biotechnology, e.g. products from genetically engineered bacteria

in natural cycles, e.g. carbon cycle (decomposers), nitrogen cycle (nitrogen-fixing bacteria, nitrifying bacteria, etc.)

Problems   – disease – named examples of plant / animal / human diseases

decay / food spoilage

Ending   – consider the good and bad points. Would we be better off without them?

5. **'Give an account of the structure and importance of nucleic acids.'**

This is probably the easiest essay on this paper. If you know your facts, you should gain high marks without too much difficulty.

Introduction – DNA / RNA found in all organisms, clearly vital to life

Components – basic structure of nucleotides, elements present, etc.

DNA   – structure, size / shape of molecule, types of bases, role in carrying the genetic code,

replication

RNA   – three types; messenger, transfer, ribosomal

description of the structure and roles of each type

comparison with structure of DNA

Importance – transfer of hereditary information from generation to generation

vital for protein synthesis

Ending   – brief description of the importance of proteins to cells.

6. **How has the work of each of the following scientists increased our undertanding of biology? (a) Mendel, (b) Darwin, (c) Watson and Crick, (d) Jacob and Monod.'**

To gain high marks in this question, there are certain key pieces of information you should include:

■ the dates they were working / published (this helps to put their work into context)

■ an outline of their famous experiment / breakthrough with reasonable scientific detail

■ how they analysed their findings / the conclusions they drew

■ how this helps our understanding today

Mendel (1865)

Mendel carried out ten years of experiments on inheritance, working with pea plants; systematically crossed parent plants, and analysed the ratios of the offspring; developed the idea of haploid gametes and diploid somatic cells; dominant and recessive alleles developed a model for monohybrid and dihybrid crosses; put forward the laws of segregation and independent assortment We still use Mendel's model to describe and predict the outcome of crosses today – it is fundamental to our understanding of genetics (when Mendel was alive, chromosomes had not been discovered).

Darwin (1858)

Darwin was an excellent naturalist, who travelled to South America and Australia on HMS *Beagle*, as part of a scientific expedition. The variety of plants and animals he saw (particularly in the Galapagos Islands) led him to develop the theory of evolution by natural selection – give a brief outline of the main points. Although not proven, there is a lot of evidence to support this – give some examples. We can see natural selection in action today e.g. peppered moths, drug-resistant bacteria, and it helps us to understand how adaptations increase the chances of survival. (Darwin also did a lot of experiments on tropisms.)

Watson and Crick (1953)

Watson and Crick built highly complex models and analysed x-ray diffraction data to determine the structure of DNA, i.e. a double helix, with base pairing between purine and pyrimidine bases. This quickly led to the breaking of the genetic code, and a proper understanding of replication and protein synthesis. Theirs was the forerunner of all of the work which has been carried out in genetic engineering over the last 40 years.

Jacob and Monod (1961)

Jacob and Monod carried out experiments on control of protein synthesis, working with the lac operon in *E. coli* – give a brief outline of how this works. Their work on control systems allows us to understand how cells can be specialised for a particular function. It was vital to know how to control protein synthesis when transferring genes to host organisms in genetic engineering.

The key to obtaining high marks lies in explaining how they have **increased our understanding of biology**: you must relate the work of each scientist to other work being done at the time, or work done since then.

Marking your essays

Read again how essays are marked on pages iv–v.
Allow up to 17 marks for scientific content.
Allow up to 3 marks for breadth.
Allow up to 3 marks for style.
Maximum mark for each essay is 20.